BABIES AT THE ZOO

3

Cheetah Cubs

Susan H. Gray

CHERRY LAKE PRESS

Published in the United States of America by
Cherry Lake Publishing
2395 South Huron Parkway, Suite 200, Ann Arbor, MI 48104
www.cherrylakepublishing.com

Content Advisor: Dominique A. Didier, Professor of Biology, Millersville University
Reading Advisor: Marla Conn, MS, Ed, Literacy specialist, Read-Ability, Inc.

Photo Credits: ©Heiko Richter/Shutterstock.com, front cover; ©Tathoms/Shutterstock.com, 1, 2; ©Mark Beckwith/Shutterstock.com, 4; ©GUDKOV ANDREY/Shutterstock.com, 6; ©Larsek/Shutterstock.com, 8; ©Francois van Heerden/Shutterstock.com, 10; ©Wollertz/Shutterstock.com, 12; ©thoron/Shutterstock.com, 14; ©TTphoto/Shutterstock.com, 16; ©nwdph/Shutterstock.com, 18; ©George Lamson/Shutterstock.com, 20

Library of Congress Cataloging-in-Publication Data

Names: Gray, Susan Heinrichs, author.
Title: Cheetah cubs / written by Susan H. Gray.
Description: Ann Arbor, Michigan : Cherry Lake Publishing, 2020. | Series: Babies at the zoo | Includes index. | Audience: Grades K-1
Summary: "Read about cheetah cubs and how zookeepers take care of them. This level 3 guided reader book includes intriguing facts and adorable photos. Students will develop word recognition and reading skills while learning about how these baby animals learn and grow, what they eat, and how they socialize with each other. Book includes table of contents, glossary, index, author biographies, sidebars, and word list for home and school connection"— Provided by publisher.
Identifiers: LCCN 2019034084 (print) | LCCN 2019034085 (ebook) | ISBN 9781534158894 (hardcover) | ISBN 9781534161191 (paperback) | ISBN 9781534160040 (pdf) | ISBN 9781534162341 (ebook)
Subjects: LCSH: Cheetah—Infancy—Juvenile literature.
Classification: LCC QL737.C23 G716 2020 (print) | LCC QL737.C23 (ebook) | DDC 599.75/91392—dc23
LC record available at https://lccn.loc.gov/2019034084
LC ebook record available at https://lccn.loc.gov/2019034085

Cherry Lake Publishing would like to acknowledge the work of the Partnership for 21st Century Learning, a Network of Battelle for Kids. Please visit http://www.battelleforkids.org/networks/p21 for more information.

Printed in the United States of America
Corporate Graphics

Table of Contents

About the Author

Susan H. Gray has a master's degree in zoology. She has written more than 150 reference books for children and especially loves writing about animals. Susan lives in Cabot, Arkansas, with her husband, Michael, and many pets.

Big, Fast Cats

What animal is the fastest runner on earth? The cheetah! This cat has a **slender** body. It has long legs. It is built to run!

Cheetahs live in Africa and Iran. Some live in zoos. Cheetahs usually have three or four **cubs** at once. Some mothers have eight!

New babies have soft tan or gray fur. At first, their eyes are closed. Their legs are weak. They cannot eat meat. They only drink milk from their mothers.

Are you **wobbly** when you wake up in the morning?

Babies have squeaky voices. They open their eyes at about 10 days. Then they begin to walk around. Cubs are very **clumsy** at first.

Zoo Babies

At zoos, the **zookeepers** often check on new babies. They clean their **pens**. Zookeepers also make sure the cubs are healthy.

Soon, cubs need more than milk. Zookeepers bring them meat. The meat might have **vitamins** mixed in.

Growing Up

As cubs grow, their coats change. They become yellow with black spots.

Little cheetahs love to run and play. **Exercise** helps their muscles grow strong.

Where is your favorite hiding place?

Cheetahs stay with their mothers for about 18 months. Moms show them safe hiding places. They also teach their cubs to hunt.

Cheetahs are adults when they are about 2 years old. Then they can have cubs of their own like these three babies.

Those little furry babies will be hungry!

Find Out More

BOOK

Meinking, Mary. *Cheetahs.* Mendota Heights, MN: Focus Readers, 2017.

WEBSITE

National Geographic Kids—Cheetah

https://kids.nationalgeographic.com/animals/cheetah/#ww-wild-cats-cheetahs.jpg

Visit this site for information about cheetahs and a map showing where they live.

Glossary

clumsy (KLUM-zee) not smooth or steady on your feet

cubs (KUBZ) young cheetahs living with their mothers

exercise (EK-sur-size) activity meant to build up the mind or body

pens (PENZ) small, closed areas for animals

slender (SLEN-dur) thin, slim

vitamins (VYE-tuh-minz) things added to food that all animals need for a healthy life

wobbly (WOB-lee) shaky or not steady

zookeepers (ZOO-kee-purz) people who take care of zoo animals

Home and School Connection

Use this list of words from the book to help your child become a better reader. Word games and writing activities can help beginning readers reinforce literacy skills.

a
about
adults
Africa
also
and
animal
are
around
as
at
babies
be
become
begin
big
black
body
bring
built
can
cannot
cat
cats
change
check
cheetah
cheetahs
clean
closed
clumsy
coats
cubs
days
drink
earth
eat
eight
exercise
eyes
fast
fastest
favorite
first
for
four
from
fur
furry
gray
grow
growing
has
have
healthy
helps
hiding
hungry
hunt
in
Iran
is
it
legs
like
little
live
long
love
make
meat
might
milk
mixed
moms
months
more
morning
mothers
muscles
need
new
of
often
old
on
once
only
open
or
own
pens
place
places
play
run
runner
safe
show
slender
soft
some
soon
spots
squeaky
stay
strong
sure
tan
teach
than
the
their
them
then
these
they
this
those
three
to
up
usually
very
vitamins
voices
wake
walk
weak
what
when
where
will
with
wobbly
years
yellow
you
your
zoo
zookeepers
zoos

Fast Facts

Habitat: Grassy plains, open forests, and deserts
Range: Parts of Africa and northeastern Iran
Length: 6 to 7 feet (183 to 213 centimeters), including the tail
Weight: 77 to 132 pounds (35 to 60 kilograms)
Life Span: 10 to 12 years
Anatomy: Cubs have a gray mane that runs down their backs. It disappears at about 3 months.
Behavior: Running at high speed uses lots of energy. So cheetahs cannot run fast for more than a minute.

Index

David Boucharc

We Learn from the Sun Teacher Lesson Plan

June 1st, 2020

ISBN - 9781989122426